HEADLINES!™

FOOD POISONING
E. COLI AND THE FOOD SUPPLY

KRISTI LEW

ROSEN
PUBLISHING®
New York

Published in 2011 by The Rosen Publishing Group, Inc.
29 East 21st Street, New York, NY 10010

Library of Congress Cataloging-in-Publication Data

Lew, Kristi.
Food poisoning: *E. coli* and the food supply / Kristi Lew. — 1st ed.
   p. cm. — (Headlines!)
Includes bibliographical references and index.
ISBN 978-1-4488-1293-6 (lib. bdg.)
1. Escherichia coli O157:H7. 2. Food poisoning—Prevention. 3. Food supply. I. Title.
QR82.E6L49 2011
615.9'54—dc22

2010020092

Manufactured in Malaysia

CPSIA Compliance Information: Batch #W11YA: For further information, contact Rosen Publishing, New York, New York, at 1-800-237-9932.

On the cover: In 2006, many grocery stores pulled bagged spinach off of shelves after an *E. coli* outbreak that was traced to the greens made dozens of people ill. The outbreak eventually spread to twenty states.

CONTENTS

Millions of people get sick every year because their food or water is contaminated with microorganisms. Microorganisms are tiny living things that include bacteria, viruses, and algae. Billions of microorganisms make their homes in our water, food, and even our bodies. Luckily, more than 95 percent of them are perfectly harmless. Indeed, some microorganisms are absolutely necessary for a healthy body. However, the other 5 percent can make people ill. In some cases, they can even be deadly.

Escherichia coli, or *E. coli*, is just such a microorganism. *E. coli* is a bacterium necessary for our health. However, it is one that can also be extremely dangerous. There are hundreds of different types, or strains, of *E. coli*. Most strains are harmless as long as they stay in their natural environment. This environment is the large intestine, or colon, of humans and other warm-blooded animals. Here, different strains of *E. coli* have a very important job to do. They help humans and animals break down their food.

However, not all types of *E. coli* keep people healthy. In fact, some strains cause illness, or even death. Unfortunately, outbreaks of illness caused by *E. coli* occur all too frequently. Outbreaks may affect only a small, isolated area, or they may affect thousands of people in multiple states or countries.

Every year millions of people around the globe get sick because their food or drinking water has been contaminated with microorganisms such as *E. coli*.

E. COLI O157:H7

The type of *E. coli* that usually makes headlines in the United States is one called *E. coli* O157:H7. The designation "O157:H7" refers to a series of chemical compounds, called markers, which are found on the surface of the bacterium. These chemical markers differentiate this strain, or serotype, from other types of *E. coli*.

E. coli O157:H7 makes people sick by producing a toxin, or poison, called Shiga toxin. Shiga toxin is extremely powerful and dangerous. It is so dangerous that the Centers for Disease Control and Prevention (CDC) classify the compound as a potential bioterrorism agent. The toxin was named for Kiyoshi Shiga, who discovered a bacterium called *Shigella* that causes dysentery, an intestinal illness. Children, the elderly, and people with weakened immune systems have the highest risk of developing serious illnesses if exposed to Shiga toxin. In some cases, these illnesses prove to be fatal. The genetic information needed to produce Shiga toxin appears in *Shigella* bacteria. Somehow, this same genetic code was transferred to *E. coli*.

The CDC first identified *E. coli* O157:H7 as a pathogen, or an agent that is capable of causing sickness, in humans in 1982. An outbreak of illness was linked to contaminated ground beef. Eating undercooked ground beef, or hamburger, is the most common food-borne means of contracting *E. coli* O157:H7 infection. For the next decade, *E. coli* O157:H7 would cause sporadic outbreaks of illness. However, it was not until 1993 that the strain would receive worldwide attention.

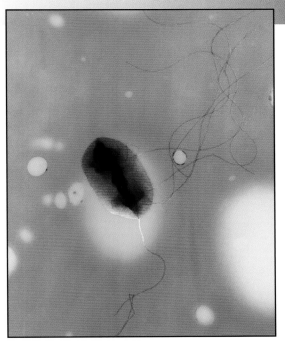

E. coli O157:H7 can make humans sick if it gets into the food or water supply. The pathogen was first identified in 1982.

OUTBREAKS

In January and February of 1993, more than 450 people in the state of Washington became ill after being infected with *E. coli* O157:H7. Forty percent of the people who became ill were under the age of six. Sixty-six percent were under the age of fifteen. The pathogen was eventually traced to undercooked hamburger at a fast-food restaurant called Jack in the Box. Not all of the people who got sick had eaten hamburger at a Jack in the Box restaurant. However, it was later determined that they had all come into contact with people who had eaten hamburger at one of the restaurants. The outbreak later moved beyond the borders of Washington. Before the outbreak was over, more than seven hundred people in four western states had become ill and four children had died.

Many *E. coli* O157:H7 infections in the United States are caused by eating hamburger that is not thoroughly cooked.

This was the biggest outbreak in the United States up to that point, but it was not the largest outbreak ever. That distinction belongs to a 1996 outbreak in Japan in which thousands of Japanese schoolchildren were infected. This outbreak did not arise from people eating ground beef. This time, contaminated radish sprouts in the children's school lunches carried the bacteria. By the end of this outbreak, more than nine thousand illnesses and two deaths were reported.

TYPES OF *E. COLI*

E. coli O157:H7 is not the only type of *E. coli* that causes people to become ill. It is not even the only type of *E. coli* that produces Shiga toxin. However, it is the most common type of Shiga-toxin-producing *E. coli,* or STEC, in the United States.

E. coli O157:H7 is a particularly virulent, or infectious, strain of food-borne bacteria. The minimum number of bacteria needed to make someone ill is called the minimum infectious dose (MID). Compared to other bacteria that cause food-borne illnesses, the MID for the O157:H7 serotype is very low. For example, *Salmonella* is another food-borne bacteria that can cause illness. Per gram of food, about one hundred thousand *Salmonella* bacteria are required to make someone ill. In

SPREADING THE SHIGA TOXIN

Scientists are not exactly sure how the Shiga toxin appeared in *E. coli*. One of the more likely scenarios involves a bacteriophage. A bacteriophage (also called a phage) is a virus that can infect a bacterium. Phages attach to specific chemicals on the surface of a bacterium. During infection, the bacteriophage injects its deoxyribonucleic acid (DNA), or genetic information, into the bacterium. The virus DNA then tells the bacterium to replicate the virus. Occasionally, some of the bacteria's genetic information gets incorporated into the virus's DNA. Eventually, the bacterium makes so many new viruses that it bursts, releasing the new viruses into the environment. If a bacteriophage did indeed infect a *Shigella* bacterium, the gene necessary to produce the Shiga toxin could have been picked up by the virus's DNA. If a virus carrying that gene later infected an otherwise harmless *E. coli* bacterium, the gene could be transferred, transforming the *E. coli* into a Shiga-toxin-producing *E. coli* (STEC) bacterium.

the same amount of food, a few dozen *E. coli* O157:H7 bacteria are enough to cause illness.

Other Shiga-toxin-producing *E. coli* strains are often lumped together and called non-O157 *E. coli*. Non-O157 bacteria also cause illness. Common non-O157 serotypes that cause illness in the United States are O26:H11, O111:NM, and O103:H2. However, outbreaks caused by these bacteria are not identified as often as those caused by O157:H7. This is mainly because it is easier for laboratories to test for the O157:H7 serotype than for non-O157 serotypes.

There are also some types of *E. coli* that may cause illness even though they do not produce Shiga toxin. Most instances of traveler's diarrhea, for example, are caused by a non-Shiga-toxin-producing *E. coli* (non-STEC). Non-STEC is also believed to be a major cause of diarrheal illnesses in developing countries, especially in children. *E. coli* strains that cause such illnesses are called enterotoxigenic *E. coli*, or ETEC. ETEC bacteria also produce a toxin, but it is a different toxin than the one STEC produces. The toxin produced by ETEC causes the lining of the intestines to produce excess fluid.

E. COLI IN DEVELOPING NATIONS

In developing nations, where proper sanitation and clean water sources may be lacking, *E. coli* is suspected of being the most common cause of childhood diarrhea. In these countries, ETEC bacteria are the most common intestinal pathogen isolated in children under the age of five. It is believed that these bacteria are responsible for several hundred million cases of illness and tens of thousands of deaths each year.

Many developing nations do not have the funding to test for bacteria routinely. In addition, some patients may not have symptoms severe enough to send them to the doctor. This makes it difficult to determine exactly how widespread illnesses caused by *E. coli* are.

How Common Is *E. Coli* Infection?

International travelers and children in developing nations are the most common victims of ETEC infection. In the United States, however, most *E. coli*–related illnesses are caused by the ingestion of food-borne *E. coli* O157:H7. Scientists estimate that about seventy thousand Americans become ill with an *E. coli* O157:H7 illness each year. About 3,500 to 7,000 of these patients develop life-threatening complications from the infection, and approximately 60 people die. Around the same number of people fall ill after being exposed to a non-O157 STEC each year as well. However, these are only estimates. Many people who experience an upset stomach, but not the more severe symptoms of a STEC infection, do not go to the doctor. As a result, their illnesses are never reported.

Since the first major outbreak in 1993, other STEC outbreaks have occurred. In the fall of 2006, an *E. coli* O157:H7 outbreak was traced to fresh spinach. According to the CDC, by the time the outbreak ended almost two hundred people from twenty-six different states had become ill after eating the contaminated greens. One hundred and two of these people were hospitalized. Thirty-one of them developed complications, and three people died as a result of their infections.

In an effort to contain the outbreak, the CDC advised consumers to cook spinach instead of eating it raw, since heat can kill the *E. coli*

In 2006, dozens of people got sick after eating spinach contaminated with *E. coli* O157:H7 bacteria. To protect their customers, grocery stores pulled the tainted greens from the shelves.

O157:H7 bacteria. The U.S. Food and Drug Administration (FDA) and the company that processed the spinach also pulled suspect packages of the vegetable from grocery store shelves. It was never determined whether the spinach had been contaminated in the field or during processing.

AN INCREASE IN CASES

The number of reported *E. coli*–related illnesses often goes up in the summer months. Bacteria, including *E. coli*, thrive in warm weather. People also tend to eat more ground beef during the summer.

The number and type of food items contaminated with *E. coli* appear to be on the rise. Between 2004 and 2006, there were approximately twenty recalls of ground beef. From January 2007 to December 2009, that number more than doubled to fifty-two. Outbreaks have not been limited to ground beef and spinach, however. In the summer of 2009, *E. coli* O157:H7 was also found in prepackaged cookie dough, which was also recalled.

CHAPTER 2

E. COLI AND THE FOOD SUPPLY

A nyone can be infected with *E. coli* if he or she ingests food or water that contains the bacteria. However, very young children, older adults, and people with weakened immune systems are the most likely to develop serious complications from the infection.

The most common symptoms of an STEC infection include severe stomach cramps, diarrhea, and vomiting. One of the symptoms that sets an STEC infection apart from an ETEC infection is bloody stool. This symptom develops because the Shiga toxin attacks the cells that line the colon. Occasionally, an infected person may also have a fever. Fevers associated with an *E. coli* infection are usually fairly low—less than 101 degrees Fahrenheit (38.3 degrees Celsius). Symptoms typically go away after five to seven days.

COMPLICATIONS

Between 5 and 10 percent of patients infected with *E. coli* O157:H7 develop complications. Some of these complications can be life threatening. The most common of these complications is a serious illness called hemolytic uremic syndrome (HUS). When someone develops HUS, the toxins produced by the *E. coli* bacteria attack the person's blood cells, destroying them.

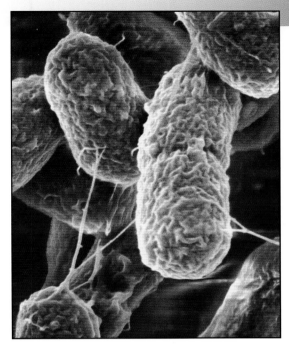

STEC infections usually come from eating food or drinking water that contains the bacteria. People infected with STEC can develop life-threatening complications.

Because their red blood cells are broken down, people suffering from HUS acquire a condition called anemia (low red blood cell count). Anemia is often indicated by extreme tiredness and a very pale skin tone. The toxin also breaks down a person's platelets, which are necessary for blood clotting. The low platelet count may cause symptoms such as unexplained bruising or bleeding from the nose or mouth.

About half of the people who develop HUS experience kidney damage and failure. Symptoms of kidney failure include a decrease in urination and the swelling of the hands, feet, face, or even the entire body. In some severe cases of HUS, seizures may also develop. Anyone experiencing symptoms of HUS should seek medical attention immediately. In case of kidney failure, the patient may need to go on kidney dialysis. Dialysis is a process that filters waste and extra fluid out of the blood. In most people with HUS, dialysis is a temporary treatment that will be stopped when their kidneys begin to function again. In some cases, however, kidney failure can be fatal.

Most people who develop HUS do eventually recover, but it may take several weeks or more. Some people experience permanent physical damage due to the infection. Although it is not common, people can die of HUS.

A *New York Times* article by Michael Moss profiled Stephanie Smith, a young woman who had life-changing problems stemming from an *E. coli* O157:H7 infection. In 2007, Smith unknowingly ate hamburger meat contaminated with the bacteria. When she started having stomach cramps and diarrhea, she thought they were caused by a simple stomach virus. However, her symptoms got much worse. The diarrhea turned bloody, her kidneys stopped working, and she started having convulsions.

In very rare cases like Smith's, the toxin that the bacteria produces can seep through the walls of the large intestine and damage blood vessels. Blood clots can develop in the damaged vessels and cause seizures. Smith's doctors were unable to stop her worsening convulsions, so they put her into a medically induced coma. The coma lasted nine weeks. When she regained consciousness, Smith's convulsions had stopped, but the toxins had damaged her nervous system. She was paralyzed from the waist down and could not walk.

While Smith's reaction to the bacteria was unusually severe, approximately 3,500 to 7,000 Americans infected with *E. coli* O157:H7 develop HUS each year. About one in three of these patients will have lingering kidney problems. Around 8 percent will develop lifelong health complications such as high blood pressure, blindness, or paralysis. A small percentage of the patients that develop HUS will die.

SYMPTOMS AND DIAGNOSIS

The amount of time that elapses between ingesting a pathogen and the appearance of the first symptoms is called the incubation time. The average incubation time for *E. coli* O157:H7 is three to four days. However, some people begin to feel ill after only a day. Others do not experience the onset of symptoms for as long as ten days. A mild stomachache is usually the first sign that something is not quite right. Over time, the stomach cramps worsen and diarrhea develops. If a person is going to develop HUS,

PERSON-TO-PERSON TRANSMISSION

Not all *E. coli* infections are food-borne. An infection can also be passed from person to person. If someone has symptoms of an STEC illness, the only way to be sure is for the person to be tested for STEC. If the person has a Shiga-toxin-producing *E. coli* infection, STEC will almost certainly be present in the stool. When the person no longer has symptoms, he or she is usually not contagious any longer. However, sometimes, especially in young children, infectious STEC bacteria can still be shed in feces for several weeks or months after symptoms disappear.

Scientists have also found that some people can be asymptomatic carriers. Asymptomatic means that a person is carrying the bacteria—and spreading it—even though he or she never develops any symptoms.

In any of these situations, the bacteria can be spread from the stool to other people due to poor hygiene. Thorough hand washing after changing diapers or using the toilet can cut down on the spread of infection.

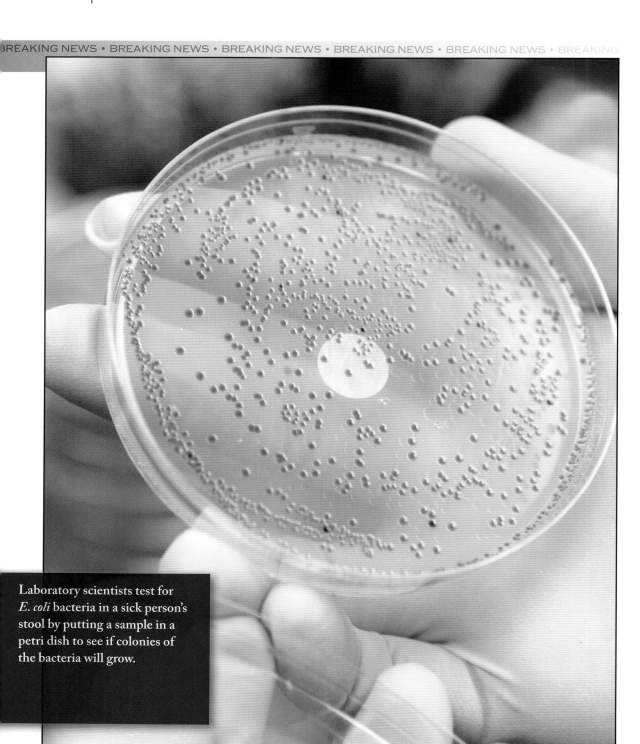

Laboratory scientists test for *E. coli* bacteria in a sick person's stool by putting a sample in a petri dish to see if colonies of the bacteria will grow.

those symptoms usually appear after the diarrhea begins to lessen, about seven days after exposure to the bacteria.

The CDC recommends that doctors test any patient suffering from bloody diarrhea for an *E. coli* infection. Doctors do this by taking a sample of the sick person's stool. The sample is sent to a laboratory that can test for the presence of *E. coli* O157:H7. During testing, the laboratory determines part of the bacteria's DNA sequence. Using this DNA "fingerprint," laboratory personnel can test other suspected cases of *E. coli* poisoning to determine if the same strain of *E. coli* is responsible. This determination helps health officials decide if there is an outbreak of *E. coli* infections originating from one source or if they are looking at random infections.

Most medical laboratories can test for *E. coli* O157:H7. They can also tell if a non-O157 STEC is present. However, some laboratories may not be able to determine the exact strain. To determine the type of non-O157 STEC making the person sick, the laboratory may send a sample to a state public health laboratory. Most of these laboratories can determine which *E. coli* serotype is causing the illness.

If a doctor believes a patient has developed HUS, he or she may request more tests. A blood test can determine if the patient is suffering from anemia or from a low platelet count. Additional tests include a urine sample to test for blood in the urine and blood tests to determine if there is any kidney damage.

Treatment and Recovery

Most people infected with STEC (who do not develop HUS) recover on their own. Patients need to drink plenty of liquids to replace the

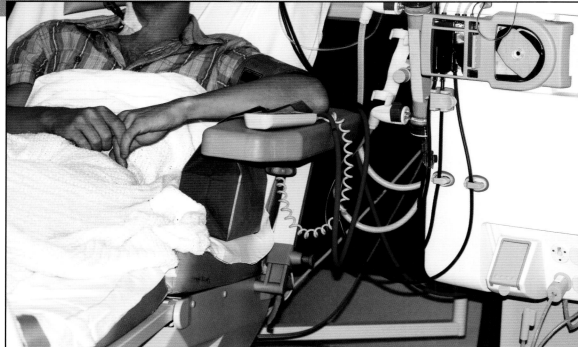

If a patient's kidneys fail due to HUS, the patient may need to undergo dialysis. A dialysis machine can temporarily do the kidneys' job of filtering waste and excess water out of the blood.

fluid they are losing from vomiting and diarrhea. They should also get a lot of rest. There are no specific medications that doctors routinely prescribe for the illness. In fact, doctors warn that anyone who suspects that he or she may have an STEC infection should refrain from taking over-the-counter antidiarrheal medications. Studies have shown that these drugs may increase the risk of developing HUS.

Upon hearing that *E. coli* is a bacterial infection, some people may conclude that their doctor should prescribe an antibiotic. Antibiotics usually kill bacteria. However, there is no scientific evidence that antibiotics work to combat STEC. In fact, they may actually increase the risk of developing HUS. Antidiarrheal medication and antibiotics may be able to help someone suffering from a non-Shiga-toxin-producing *E. coli* illness, however.

It is, therefore, important for anyone who believes that he or she may be suffering from an *E. coli*–related illness to see a doctor. If the diarrhea is caused by a non-STEC, the doctor may be able to help. It is also important for doctors to know about any STEC-related illnesses. This is so they can monitor the patients for complications and report these illnesses to help protect the public's health. Medical personnel (doctors and laboratory technologists, for example) are required to notify the CDC of any STEC illnesses they encounter. This reporting system allows CDC scientists to determine if they are dealing with isolated cases or if an *E. coli* outbreak is developing.

Although most people who develop an *E. coli*–related illness will get better whether they see a doctor or not, people with HUS require medical attention. Treatment for HUS may include blood transfusions. A patient with HUS may also require kidney dialysis. In severe cases, the kidneys may fail permanently. If this happens, a kidney transplant is necessary.

MODERN OUTBREAKS

E. *coli* makes people sick when it goes where it does not belong, such as in people's food or water supply. Non-Shiga-toxin-producing *E. coli* has been making people sick for a very long time—much longer than *E. coli* O157:H7. However, there has been an increase in *E. coli* O157:H7–related illnesses in the last few decades.

WHY ARE THERE MORE OUTBREAKS?

Several factors have converged to cause this increase. First, the emergence of Shiga-toxin-producing *E. coli* O157:H7 is relatively new. Before 1982, this bacteria was unknown. Unlike other strains of *E. coli* that must be ingested in significant amounts to cause illness, even small amounts of *E. coli* O157:H7 can cause serious illness.

Another factor contributing to the increase is that this Shiga-toxin-producing serotype has flourished in the intestinal tracts of cattle, camels, and other ruminant (plant-eating, cud-chewing) animals. The bacteria usually does not make these animals sick, but it can make humans ill if they come into contact with it.

Pathogenic *E. coli* gets into the environment when these plant-eating animals defecate. Once the STEC is in the environment, other

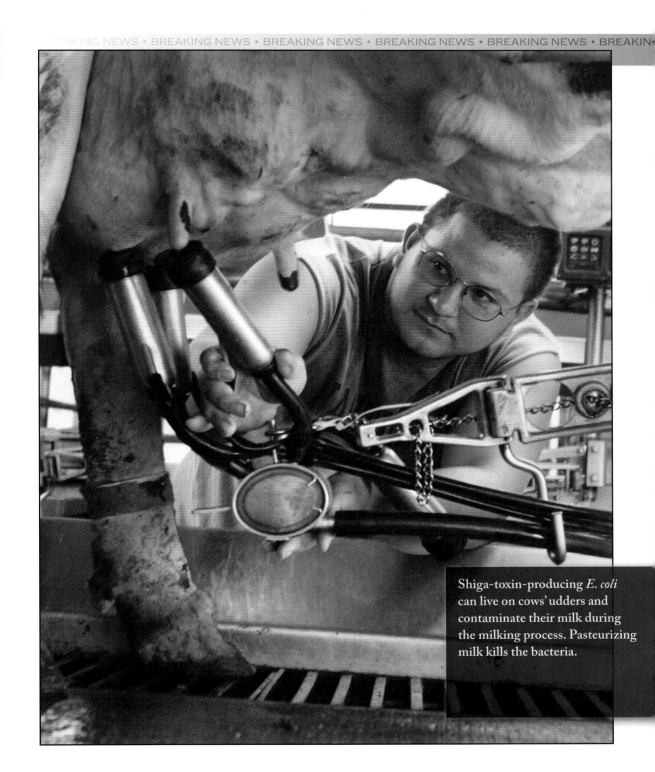

Shiga-toxin-producing *E. coli* can live on cows' udders and contaminate their milk during the milking process. Pasteurizing milk kills the bacteria.

animals, such as pigs or birds, may come into contact with the bacteria and spread it. Infected feces can also get into the water supply, contaminating it. Finally, *E. coli* bacteria can enter the food system when animals are processed in slaughterhouses. If the contents of the animals' intestines come into contact with the meat, the meat can become contaminated.

Humans can become infected with *E. coli* when they ingest the bacteria. Shiga-toxin-producing *E. coli* can live on a cow's udder, for example. When the cow is milked, STEC can contaminate the milk. Most of the milk found in grocery stores, however, has undergone a process called pasteurization to kill the bacteria. Unpasteurized milk, and cheese made from unpasteurized milk, has a high risk of being contaminated with *E. coli* O157:H7. So does unpasteurized apple juice. People may also get sick if they swallow contaminated water while swimming, eat food prepared by those who did not wash their hands after using the bathroom, eat vegetables grown in dirt that is contaminated with animal feces, or otherwise come into contact with feces containing STEC.

In the United States, the most common avenue of STEC infection is processed food, mainly ground beef. The way American families get their food has also been a factor in the large, multistate STEC outbreaks. Many years ago, families were much more likely to buy all of their meat, milk, and vegetables from a local farmer. If these food items were contaminated with *E. coli*, any illnesses that resulted would be limited to a relatively small area. Today, Americans get their food from all over the world. Food that is grown in one country can be easily transported to many others. Contaminated meat, spinach, or other food items can reach thousands of people.

Ground Beef

Food processing methods contribute to the spread of *E. coli*. This is especially true for ground beef. The meat in a package of ground beef purchased in an American grocery store is not taken from a single cow. Instead, hamburger meat is made from the trimmings of different cuts of beef from a number of different cows. The trimmings arrive at the meat grinders from many different slaughterhouses. This practice, which is widespread in the meat grinding industry, allows grinders to produce hamburger with the desired amount of fat for the least amount of money. When cows are being slaughtered and their meat is being processed, matter from inside their intestines, which contain *E. coli*, sometimes comes in contact with the meat, contaminating it.

The beef patties that Stephanie Smith ingested, for example, contained ingredients from slaughterhouses and meat processing plants in Nebraska, Texas, South Dakota, and Uruguay. Cargill, the company that made the beef patties, uses trimmings from more than fifty different vendors. Each one of these companies has its own rules when it comes to testing for *E. coli* in the beef.

Food experts believe the trimmings used to make ground beef are even more susceptible to *E. coli* infection than other cuts of beef. The cuts of meat used for ground beef are usually taken from the surface of the carcass, which is more likely to get smeared with feces. Even if

Cuts of beef from several different animals are used to make ground beef. If any of these cuts have STEC on them, the entire batch of hamburger will be tainted.

grinders used cuts from interior portions of the carcass, it would not guarantee that the meat would be free of *E. coli*. If the animal's intestines are accidentally cut open during the slaughtering process, the meat can become contaminated.

The trimmings are sent to a meat grinder to be turned into ground beef. If any of the trimmings have *E. coli* on them, the bacteria gets thoroughly mixed in with the ground beef.

TREATING BEEF WITH AMMONIA

About 10 percent of the meat in Smith's burger patty was supplied to Cargill by a South Dakota company called Beef Products, Incorporated (BPI). This company buys high-fat trimmings that were once considered useless and thrown away. BPI heats the trimmings to liquify the fat. The fat is then skimmed off the top. The resulting meat has less fat and is used in many beef products, including fresh and frozen hamburger patties, low-fat hot dogs, and luncheon meats.

To prevent *E. coli* contamination, BPI treats processed meat with ammonia. Ammonia, a common cleaning agent, is an antimicrobial— a substance that kills or slows the growth of microorganisms. According to the *New York Times*, a study conducted by Iowa State University (and paid for by BPI) determined that treating beef with ammonia reduces *E. coli* bacteria to undetectable levels. Based on this research, the U.S. Department of Agriculture (USDA) decided that the process was a safe and effective method of decreasing the amount of *E. coli* in meat. The USDA also decided to exempt BPI from its *E. coli* contamination testing program, which began in 2007.

Voluntary Testing

Some slaughterhouses test for *E. coli* O157:H7 before their meat is sent to the grinders, but others do not. Testing for the bacteria is strictly voluntary and is not enforced by law. Most meat grinders do not do their own testing for *E. coli*. There are a few exceptions, such as Costco, which has its own grinding facility. This company insists that grinders test the meat for *E. coli* contamination before it is ground up and offered for sale to the public.

Major fast-food restaurants, such as McDonald's and Burger King, use BPI's trimmings as one of the ingredients in their hamburgers. Along with providing ground beef to fast-food restaurants and grocery stores, BPI also sells its beef to the federal government for use in school lunch programs. In 2006, 2008, and 2009, the federal school lunch program found *E. coli* contamination in BPI trimmings, despite the fact that the beef had been treated with ammonia. Because of the testing, none of the contaminated meat reached the schools, but it did raise questions about the effectiveness of the ammonia treatment process.

After contamination was again detected in 2009, the USDA decided that BPI could no longer be exempt from the department's routine *E. coli* testing. BPI acknowledged that it was treating the beef with less ammonia. Company officials said they changed the process because they were receiving complaints about the smell and taste of the treated beef. BPI insists that the process still works, even with the lowered amount of ammonia. It has supplied the USDA with an additional study, which is still under review.

BPI is not the only company from which Cargill bought beef to make the patties that made Stephanie Smith ill. Slaughterhouses in Nebraska, Texas, and Uruguay also contributed beef to that batch of meat. Because Cargill does not test the trimmings that it grinds for *E. coli*, it is impossible to tell which, if any, of these slaughterhouses was the source of the bacteria.

The Nebraska company Greater Omaha Packing reports that it washes its animal carcasses in hot water to remove any traces of feces that might carry *E. coli*. However, the Texas company Lone Star Beef Processors told the *New York Times* that even the cleanest slaughterhouses cannot guarantee that trimmings will be completely free of the bacteria. A USDA inspection of the slaughterhouse in Uruguay revealed sanitation problems and deficiencies in the way the slaughterhouse conducted *E. coli* testing. The company has said that it has corrected these problems.

The U.S. government banned the sale of ground beef known to be contaminated by *E. coli* O157:H7 after the 1993 Jack in the Box outbreak. However, because of nonstandardized testing, contamination is not always caught. Federal health officials estimate that tens of thousands of Americans get sick from eating food contaminated with *E. coli* O157:H7 every year. Between 2006 and 2009, ground beef was identified as the source in sixteen outbreaks.

FRUITS, VEGETABLES, AND WATER

While most *E. coli*–related illnesses in the United States stem from tainted ground beef, other food items, such as spinach, lettuce, and prepackaged cookie dough, have also been implicated

Water that has come into contact with human or livestock waste may contain *E. coli*. Drinking contaminated water can make people very ill.

in outbreaks. Fruit and vegetable crops can become contaminated if they are irrigated with water that has come into contact with human sewage or livestock feces. Infected cow manure can also contaminate fruits and vegetables if it is used to fertilize the crops. It is also possible for produce to become contaminated during processing, possibly through rodent droppings. Washing produce that has come into contact with *E. coli* is usually not enough to prevent an infection. Cooking produce thoroughly, however, is usually effective.

CHAPTER 4

IS OUR FOOD SUPPLY SYSTEM FLAWED?

The main strategy used by food manufacturers and the U.S. government to prevent the spread of *E. coli* today is product recall. In 2009, more than 2 million pounds (907,184 kilograms) of beef were recalled after contamination was discovered. In the first two months of 2010, recall amounts already topped 500,000 pounds (226,796 kg) of beef. Unfortunately, food products are typically only recalled after illness has already been reported.

On October 31, 2009, 545,699 pounds (247,524 kg) of ground beef was recalled because of *E. coli* contamination. Donna Rosenbaum, executive director of Safe Tables Our Priority (S.T.O.P.), an organization that works to ensure a safe food supply, told the *New York Times* that the recall showed that our country's food testing system is not working. This time, the deadly bacteria was traced to an Ashville, New York, beef processing plant owned by AFA Foods. Rosenbaum and many others believe that contaminated meat should be detected before it leaves a company's warehouses, not after people have fallen ill. However, AFA Foods told the *New York Times* that it was following the beef industry's testing guidelines.

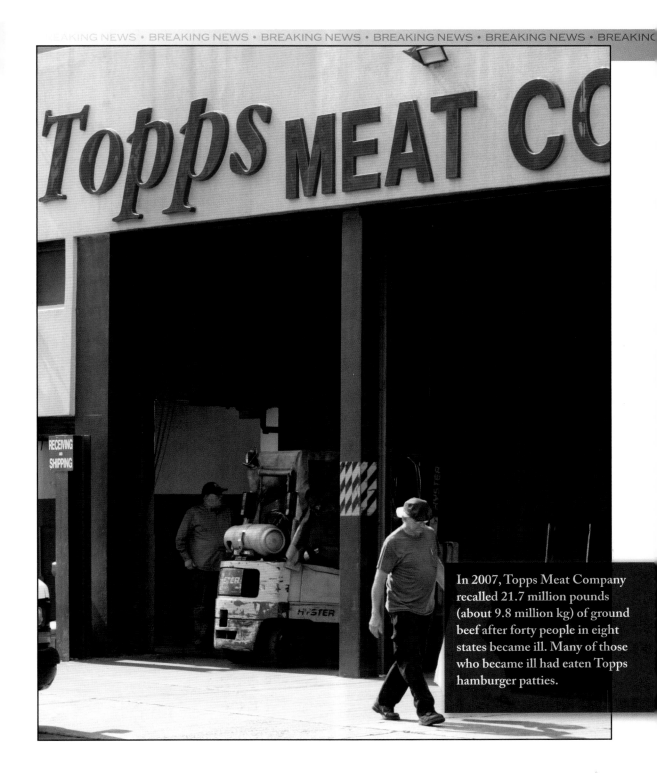

In 2007, Topps Meat Company recalled 21.7 million pounds (about 9.8 million kg) of ground beef after forty people in eight states became ill. Many of those who became ill had eaten Topps hamburger patties.

The slaughterhouses from which AFA Foods buys its beef test the trimmings for *E. coli*. A spokesperson from AFA Foods reported that AFA also tests samples of its ground beef, sometimes as often as every ten minutes. The company's spokesperson reported that, during the time the recalled beef was manufactured, all tests for *E. coli* bacteria came back negative. Indeed, if *E. coli* had been found, it would have been against the law for the company to sell the contaminated meat.

However, not every batch of ground meat is tested. Some contaminated beef slipped through. Before the beef was recalled, hundreds of people had become ill and two had died of *E. coli* poisoning. Near the end of 2009, AFA Foods was actively looking into changing their testing protocol.

CHANGING THE SYSTEM

The testing protocol for ground beef could be made more rigorous. For instance, grinders could test meat trimmings from the slaughterhouse and reject any that contain *E. coli* bacteria. As the system stands at present, most grinders do not do this. Slaughterhouses and some meat industry trade groups have been resistant to independent testing by grinders. They claim that independent testing is unnecessary, as slaughterhouses already test the trimmings. These groups believe that testing would be redundant and only serve to increase cost. They are also concerned that, because slaughterhouses sell to many different grinders, additional testing would lead to larger and more frequent recalls. For example, if one grinder discovers a batch of trimmings contaminated by *E. coli*, other companies that have bought trimmings from the same slaughterhouse might return the meat they purchased.

Many slaughterhouses and grinders employ scientists to test beef for the presence of *E. coli* O157:H7. However, the testing is not mandatory, and the bacteria is not always detected.

Officials from AFA Foods confirmed that, at one time, they tested trimmings coming into their facilities. However, they stopped when the slaughterhouses they worked with objected to this practice and threatened not to sell to them.

For now, AFA Foods and most other grinders only test for *E. coli* once all of the trimmings have been ground together. This method has its problems—a fact that AFA Foods found out in 2007 when a batch of ground beef tested positive for *E. coli*. Because the company uses many different suppliers, there was no way to determine which slaughterhouse had sold the company the contaminated trimmings. Without this knowledge, there was also no way to warn other companies that may have purchased meat from that slaughterhouse. This is the same situation Cargill experienced during the outbreak that made Stephanie Smith sick.

In his *New York Times* article about Smith, Michael Moss reported that a test of the independent laboratories on which the food industry relies revealed that the labs are not always accurate. In this test, samples of ground beef were deliberately contaminated with *E. coli* and sent to various testing laboratories. The results showed that some laboratories failed to find contamination in up to 80 percent of the samples.

Beginning in 2007, the USDA began testing for *E. coli* O157:H7 as well. However, the department is only able to test approximately fifteen thousand batches of meat a year. These spot checks are carried out in meat processing plants and grocery stores all over the country.

ANTIBIOTIC-RESISTANT *E. COLI*

Scientists are concerned about the emergence of some types of antibiotic-resistant *E. coli*. Antibiotic-resistant bacterial infections can be extremely difficult to treat. This is especially true if the bacteria have become resistant to more than one type of antibiotic. These are called multidrug-resistant (MDR) bacteria.

Near the end of 2009, a group of scientists at the University of Calgary reported the sudden emergence of a strain of *E. coli* that causes urinary tract infections (UTIs). This strain of the bacteria has developed the ability to produce an enzyme that breaks down the two most common antibiotics used to treat UTIs. The enzyme renders these drugs useless.

The emergence of MDR–*E. coli* doesn't just make bacterial infections more difficult and expensive to treat—it may also increase the number of *E. coli* O157:H7 outbreaks. Robert Lawrence, director of the Johns Hopkins Center for a Livable Future, and many other health experts believe that the way farmers feed their livestock is contributing to the problem of antibiotic-resistant bacteria. Sick animals need antibiotics to make them healthy again. However, curing sick animals is not the only way antibiotics are used on today's farms. Many healthy animals are fed antibiotics not to treat an illness, but to make them grow faster. The faster an animal grows, the more quickly it can be sent to market. This approach decreases the amount of time an animal needs to be fed. It also increases the number of animals that can be brought to market each year.

The problem with using antibiotics in this manner is that the more often *E. coli* (or any other bacteria) encounters a particular antibiotic, the more likely it is that the bacteria will develop resistance to it. At first, an antibiotic will kill off many of the bacteria it is targeting.

THE SPREAD OF ANTIBIOTIC-RESISTANT GENES

Some experts are concerned that the antibiotic-resistant genes in one bacterial species may spread to other types of bacteria through a genetic vector. A genetic vector is a molecule that can pass genetic material from one cell to another. A genetic vector—specifically, a bacteriophage—is the way scientists believe that the Shiga toxin got into *E. coli* bacteria to begin with. For these reasons, scientists strongly advise that farmers stop giving animals antibiotics for non-therapeutic reasons.

After the *E. coli* outbreak linked to Jack in the Box restaurants sickened and killed children in 1993, parents and families formed a food safety organization called Safe Tables Our Priority (S.T.O.P.).

However, some of the bacteria may naturally carry genes that make them resistant to that antibiotic. The drug kills off the bacteria that do not have the resistant genes, leaving behind the ones that do. The surviving bacteria reproduce and pass on the antibiotic resistance to the next generation. Animals can then shed the antibiotic-resistant bacteria in their feces, possibly infecting other animals or contaminating water.

CHAPTER 5

PREVENTING THE SPREAD OF *E. COLI* INFECTIONS

I n the first few months of 2010, more than 500,000 pounds (226,796 kg) of ground beef were recalled because of *E. coli* contamination. Complications from *E. coli* O157:H7 infections are now the leading cause of kidney failure in American children. Given these facts, what can people do to protect themselves and prevent the spread of *E. coli* bacteria?

MANDATED TESTING

If Senator Kirsten Gillibrand of New York has anything to say about it, increased testing within the meat processing industry will become mandatory. Gillibrand believes that meat companies and the USDA are not currently doing everything within their power to make sure that consumers are as safe as they could be. To remedy this situation, Gillibrand has proposed legislation that would require companies to test meat for *E. coli*.

Gillibrand introduced the *E. coli* Eradication Act of 2009 to Congress on November 18, 2009. In order for the bill to become law, it must first pass through a congressional committee. Congressional committees are groups of lawmakers that study a problem and the proposed solution. As of January 2010, the bill was still in the hands of the Committee on Agriculture, Nutrition, and Forestry. If the bill makes it out of committee and is signed into law, it would require slaughterhouses to test the meat that leaves their facilities to ensure that no *E. coli* bacteria is present. In addition, the bill would require grinders to retest the trimmings before mixing them together.

The American Meat Institute does not believe legislation such as the *E. coli* Eradication Act of 2009 is necessary. The organization notes that slaughterhouses already do their own voluntary spot checks and argues that increased testing would mean increased costs. Mandatory testing would, meat industry officials say, put an unfair financial burden on small meat manufacturing plants. A spokesperson from Senator Gillibrand's office, however, argues that the price of increased testing would be about a penny per pound (.45 kg) of ground beef.

CATTLE VACCINATIONS

Another idea for controlling *E. coli* O157:H7 infections in humans is to attempt to control the amount of the bacteria in cattle. Two companies have heeded the call to try to vaccinate away the troublesome bacteria. In March 2009, Epitopix, a Minnesota-based company, received USDA approval to sell its vaccine while scientists continue to research it. The company planned to test the vaccine on approximately three hundred thousand cattle in early 2010. The Canadian company

Bioniche Life Sciences also has an *E. coli* vaccine. It is approved for use in Canada, but not yet for use in the United States.

Preliminary tests of the Epitopix vaccine have shown that 86 percent of the cattle vaccinated no longer shed *E. coli* bacteria in their feces. The remaining 14 percent still had traces of *E. coli* in their waste, but the number of bacteria was cut by 98 percent. Both of these results mean a decreased number of bacteria in the environment, which should lead to fewer infected animals.

The vaccines from both Epitopix and Bioniche have actually been available for several years. However, there have been delays in getting them approved. The USDA first received a request for approval of an *E. coli* vaccine in 2001. Two years later, USDA officials determined that, by federal law, the department could not make a ruling on the vaccine. (The USDA is only allowed to make decisions on vaccines that involve animal health. But *E. coli* does not make cattle sick, only humans.) Therefore, the application was passed to the Food and Drug Administration. However, the FDA refused the application because it is responsible for human health and welfare, not approving animal vaccines.

In 2005, the USDA changed its decision and took the application for a cattle vaccine. The department decided that in order for the vaccine to be approved, studies would need to show at least a 90 percent reduction in the number of animals carrying *E. coli*. Neither vaccine could accomplish this, so neither one was approved.

It looked like a dead end for both of these drugs until 2007, when the number of *E. coli* illnesses and beef recalls increased sharply. In early 2008, the USDA announced that it would lower its

Health officials hope a new vaccine will reduce the number of *E. coli* O157:H7–infected cattle, keeping more of the bacteria out of the food supply.

requirements. About a year later, the department granted Epitopix a conditional license to sell the vaccine while its research continues.

THE COST OF VACCINATION

Even if testing proves that a vaccine works, and it is approved by the USDA, farmers and feedlot owners are concerned about the price of vaccinating cattle. The *New York Times* reports that Bioniche sells its vaccine for about $3 per dose in Canada. Complete vaccination requires two or three doses, totaling $6 to $10 per animal. Ten dollars may not sound like a lot, but Colorado feedlot owner Jason Timmerman told the *New York Times* that, in a good year, his profit per animal is only between $25 and $35 to begin with. Timmerman argues that if farmers become responsible for paying for a cattle vaccine, their profits would decline while the meat packers would be rewarded with decreased recalls and lawsuits.

For now, Cargill is paying part of the cost of vaccinating one hundred thousand heads of cattle in a study designed to test the effectiveness of the Epitopix vaccine. Testing will be conducted to see if meat and trimmings from vaccinated cattle contain less bacteria than meat from unvaccinated animals.

Scientists do not think that a cattle vaccine will be enough to completely wipe out *E. coli* O157:H7. However, it may be enough to reduce the number of bacteria entering slaughterhouses. With the number of bacteria reduced, experts hope that safety measures that are already in place, such as hot water baths and steam pasteurization, will be enough to eliminate the bacteria from the food supply.

Other Methods of Killing Bacteria

Another method that is also being tested by Cargill is spraying cattle with a substance that contains a bacteriophage. The bacteriophage that is being tested on cattle specifically targets chemicals that are only found on the surface of *E. coli*. Studies have shown that the virus will not infect humans. Also, if it cannot find an *E. coli* bacterium to infect, it will just break down. The spray being tested by Cargill is made by the animal health division of the pharmaceutical company Eli Lilly.

Currently, slaughterhouses use a combination of washing, scrubbing, rinsing, and steaming to eliminate as much *E. coli* bacteria as possible. Other methods for controlling the bacteria are being tested.

An additional method that is being tested to control *E. coli* bacteria involves a chemical called sodium chlorate. This chemical is currently used in environmentally friendly paper bleaching. In an anaerobic, or oxygen-free, environment such as a cow's gut, *E. coli* bacteria use nitrogen to produce energy. However, *E. coli* is not good at distinguishing between nitrogen and chlorate. If chlorate is present, it will try to use that instead. When sodium chlorate is given to cattle in small doses, it acts as a sort of suicide pill for *E. coli*: inside of *E. coli* bacteria, sodium chlorate becomes a sodium hypochlorite solution, or bleach. The bleach kills the bacteria without hurting the cattle. Georgia company Eka Chemicals is working on licensing the chemical for this use. The FDA is reviewing it for approval.

PERSONAL SAFETY

In the meantime, the CDC has several recommendations for how people can protect themselves from *E. coli* poisoning. One of the best ways to prevent spreading the bacteria is to wash one's hands routinely. This is particularly important after going to the bathroom, after changing a baby's diaper, and before preparing or eating food. Hands should also be washed after contact with animals, whether on a farm, in a petting zoo, or in an everyday environment.

The CDC recommends that cutting boards, knives, and other utensils that have come into contact with raw meat be washed thoroughly in hot water before being used again. Countertops should also be washed. In a warm kitchen, the number of *E. coli* bacteria will double in forty-five minutes. The bacteria can survive from a few hours to an entire day on these surfaces.

To protect yourself from *E. coli* poisoning, make sure ground beef is thoroughly cooked. The meat should reach an internal temperature of at least 160 degrees Fahrenheit (71 degrees Celsius).

Pasteurization is a process that uses heat to kill harmful bacteria in liquids, such as fruit juice or milk. Avoid drinking unpasteurized liquids or food products made with them.

Food scientists warn, however, that these guidelines may not be enough. Dr. Mansour Samadpour, a scientist from the meat industry's leading testing company, helped the *New York Times* staff carry out their own *E. coli* test. To perform the test, a sample of ground beef was purposely contaminated with a nonvirulent strain of *E. coli*. The staff then used and washed a cutting board like they would in an ordinary kitchen. Even after they washed it in hot water, tests showed that *E. coli* still remained. They also found that a kitchen towel used during the testing contained a large amount of the bacteria. Moss's conclusion was that the CDC's guidelines may not prevent the bacteria from spreading in an ordinary kitchen.

People should make sure that all of the meat they eat is thoroughly cooked. They should also avoid drinking unpasteurized apple juice and milk, and avoid eating cheeses made from unpasteurized milk. Frozen juices and the juices sold in glass bottles or boxes at room temperature in the grocery store have all been pasteurized.

Food Safety

The best way people can protect themselves from *E. coli* O157:H7 is to make sure that all of the meat they eat is thoroughly cooked. Hamburgers should reach a cooking temperature of at least 160ºF (71ºC). If a hamburger is still pink in the center, it needs to be cooked more. The color of the meat alone is not always a good indication of how well done it is. The only way to ensure that meat has reached the proper internal temperature is to use a meat thermometer.

Sometimes a community's water supply can become contaminated with *E. coli*. This may occur if a water main is broken, a natural disaster has affected the area, or some other accident has caused a breakdown in normal water sanitation. If local officials say it is necessary, boil water before drinking it or cooking with it. Drink water only from safe sources, such as tap water that has been treated with chlorine, bottled water, or well water that has been tested for contamination. Also, avoid swallowing water while swimming in lakes, ponds, or swimming pools. The chlorine in swimming pools normally kills the bacteria, but if the pool is not sufficiently chlorinated, there may be traces of *E. coli* in the water. Boiling vegetables that have possibly come into contact with contaminated water for at least one minute will kill the bacteria also. It's important to pay attention to local and national news to help protect yourself and your family. If health officials determine that there has been an STEC outbreak, they will announce which products and areas they believe to be affected. Avoid eating or drinking anything that may have become contaminated.

Public Awareness

Until testing in the meat industry is standardized, cattle vaccines are studied and approved, or other methods of decreasing *E. coli* in the environment become widespread, public awareness is the biggest weapon in the arsenal against *E. coli* contamination. Increased scrutiny of the food industry and the government departments that regulate it has led to a review of current practices. Public awareness campaigns have also made many people aware that they can protect themselves by making sure their meat is thoroughly cooked. Vigilantly checking food items that may already be in the family's refrigerator or freezer against announced recalls can also decrease the number of people affected if an outbreak does occur.

anaerobic Able to live and grow in an oxygen-free environment.

anemia A condition in which the blood is deficient in red blood cells.

antibiotic A substance that kills, weakens, or slows the growth of bacteria.

antimicrobial A substance that kills or slows the growth of microorganisms.

asymptomatic carrier A person or animal that can carry a pathogen without having any symptoms of disease.

bacteriophage A virus that infects and destroys bacteria.

colon The longest part of the large intestine.

dialysis A medical process that removes wastes or toxins and extra fluid from the blood when the kidneys cannot do so.

incubation time The amount of time between infection with a pathogen and the appearance of symptoms of the illness or disease it causes.

irrigation Artificially watering crops.

manure Plant and animal waste used to fertilize crops and other plants.

microorganism Microscopic organisms such as bacteria, viruses, algae, yeast, or protozoa.

pasteurization A process in which heat is used to kill pathogens in milk and other food products.

pathogen A disease-causing organism.

recall A request to return a product to the maker, due to the discovery of health or safety issues.

ruminant animal A plant-eating mammal that chews cud (regurgitated, partially-digested food). Ruminant animals include cattle, camels, goats, and deer.

serotype A group of microorganisms that have the same chemical compounds on their surfaces.

sewage Liquid and solid waste carried away from homes and businesses in sewers or down drains.

transfusion To transfer red blood cells or platelets from one person to another.

virulent Highly infectious and able to cause severe illness or death.

FOR MORE INFORMATION

Canadian Food Inspection Agency (CFIA)

1400 Merivale Road

Ottawa, ON K1A 0Y9

Canada

(800) 442-2342

Web site: http://www.inspection.gc.ca/english/toce.shtml

The CFIA works to make sure that the Canadian food supply is as safe and sustainable as possible. The agency's Web site offers resources for consumers, including recall lists, food safety tips, and allergy alerts.

Canadian Partnership for Consumer Food Safety Education

c/o Brenda Watson, Executive Director

R.R. #22

Cambridge, ON N3C 2V4

Canada

(519) 651-2466

Web site: http://www.canfightbac.org/cpcfse/en

The Canadian Partnership for Consumer Food Safety Education increases awareness of safe food-handling practices. The organization provides learning guides for teachers, educational brochures for families, and games and activities for kids.

Centers for Disease Control and Prevention (CDC)

1600 Clifton Road

Atlanta, GA 30333

(800) 232-4636

Web site: http://www.cdc.gov

The CDC works to protect public heath in the United States by providing general information about food-borne pathogens, and by tracking and providing updates about potential outbreaks.

International Food Information Council Foundation (IFIC)

1100 Connecticut Avenue NW, Suite 430

Washington, DC 20036

(202) 296-6540

Web site: http://www.foodinsight.org/Home.aspx

The IFIC is a nonprofit organization that educates consumers by collecting and reporting on the latest science-based health, nutrition, and food safety information.

President's Food Safety Working Group

U.S. Department of Health and Human Services

200 Independence Avenue SW

Washington, DC 20201

Web site: http://www.foodsafetyworkinggroup.gov/Home.htm

The President's Food Safety Working Group is a group of food safety advisers to the U.S. president. The group maintains a Web site that allows consumers to report food suspected of contamination, look for recalls, sign up for alerts, and ask experts about food safety.

Safe Tables Our Priority (S.T.O.P.)
3149 Dundee Road, #276
Northbrook, IL 60062
(847) 831-3032
Web site: http://www.safetables.org/index.cfm
S.T.O.P. is a nonprofit organization founded in 1993. The group supports victims of food-borne illnesses and fights for changes in public policy that affect food safety and education.

WEB SITES

Due to the changing nature of Internet links, Rosen Publishing has developed an online list of Web sites related to the subject of this book. This site is updated regularly. Please use this link to access the list:

http://www.rosenlinks.com/hls/coli

Ballard, Carol. *Food Safety* (What If We Do Nothing?). Pleasantville, NY: Gareth Stevens Publishing, 2010.

Ballard, Carol. *Is Our Food Safe?* (Global Issues). London, England: Arcturus Publishing, 2008.

Bjorklund, Ruth. *Food Borne Illnesses* (Heath Alert). New York, NY: Benchmark Books, 2006.

Casper, Julie. *Agriculture: The Food We Grow and Animals We Raise* (Natural Resources). New York, NY: Chelsea House Publications, 2007.

Guilfoile, Patrick. *Antibiotic Resistant Bacteria* (Deadly Diseases and Epidemics). New York, NY: Chelsea House Publications, 2006.

Hunter, Beatrice. *Infectious Connections: How Short-Term Foodborne Infections Can Lead to Long-Term Health Problems*. Laguna Beach, CA: Basic Health Publications, 2009.

Johanson, Paula. *Processed Food* (What's in Your Food? Recipe for Disaster). New York, NY: Rosen Publishing Group, 2008.

Kallen, Stuart. *Is Factory Farming Harming America?* Detroit, MI: Greenhaven Press, 2006.

Levete, Sarah. *Toxins in the Food Chain*. New York, NY: Crabtree Publishing Company, 2010.

Manning, Shannon. Escherichia Coli *Infections* (Deadly Diseases and Epidemics). New York, NY: Facts on File, 2004.

May, Suellen. *Invasive Microbes* (Invasive Species). New York, NY: Chelsea House Publications, 2007.

Owen, Ruth. *Growing and Eating Green: Careers in Farming, Producing, and Marketing Food*. New York, NY: Crabtree Publishing Company, 2009.

Pollan, Michael. *The Omnivore's Dilemma for Kids: The Secrets Behind What You Eat*. New York, NY: Dial Books, 2009.

Schlosser, Eric, and Charles Wilson. *Chew on This: Everything You Don't Want to Know About Fast Food*. Boston, MA: Houghton Mifflin, 2006.

Sherrow, Victoria. *Food Safety* (Point/Counterpoint). New York, NY: Facts on File, 2008.

Sieling, Peter. *Growing Up on a Farm: Responsibilities and Issues*. Broomall, PA: Mason Crest Publishers, 2008.

Silverstein, Alvin, Virginia Silverstein, and Laura Silverstein Nunn. *The Food Poisoning Update* (Disease Update). Berkeley Heights, NJ: Enslow Publishers, 2007.

Stille, Darlene. *Recipe for Disaster: The Science of Foodborne Illness* (Headline: Science). Minneapolis, MN: Compass Point Books, 2010.

Weber, Karl, ed. *Food Inc.: A Participant Guide: How Industrial Food Is Making Us Sicker, Fatter, and Poorer—and What You Can Do About It*. New York, NY: PublicAffairs, 2009.

Zimmer, Carl. *Microcosm:* E. coli *and the New Science of Life*. New York, NY: Knopf Doubleday Publishing Group, 2008.

BIBLIOGRAPHY

Betsy, Tom, and James Keogh. *Microbiology Demystified*. New York, NY: McGraw-Hill, 2005.

Centers for Disease Control and Prevention. "*E. coli*." June 19, 2009. Retrieved February 26, 2010 (http://www.cdc.gov/ecoli).

Fitzgerald, Randall. *The Hundred-Year Lie: How Food and Medicine Are Destroying Your Health*. New York, NY: Dutton, 2006.

Gever, John. "Drug-Resistant *E. coli* Called Significant Threat." MedPage Today, December 28, 2009. Retrieved February 26, 2010 (http://www.medpagetoday.com/Urology/Urology/17697).

Govtrack.us. "S. 2792: *E. coli* Eradication Act of 2009." November 18, 2009. Retrieved February 26, 2010 (http://www.govtrack.us/congress/bill.xpd?bill=s111-2792).

Harris, Gardiner. "*E. coli* Kills 2 and Sickens Many; Focus Is on Beef." *New York Times*, November 2, 2009. Retrieved February 26, 2010 (http://www.nytimes.com/2009/11/03/health/03beef.html?fta=y).

Hilts, Philip. "Gene Jumps to Spread a Toxin in Meat." *New York Times*, April 23, 1996. Retrieved February 26, 2010 (http://www.nytimes.com/1996/04/23/science/gene-jumps-to-spread-a-toxin-in-meat.html?pagewanted=1).

Hovey, Art. "Cargill's *E. coli* Vaccine Testing Draws Attention." *Lincoln Journal Star*, January 4, 2010. Retrieved February 26, 2010 (http://www.journalstar.com/news/state-and-regional/nebraska/article_35a434e2-f8c7-11de-806c-001cc4c03286.html).

Lawrence, Robert. "Dangerous Medicine in U.S. Food
    Production: Antibiotics." Johns Hopkins Center for a
    Livable Future, March 16, 2009. Retrieved February
    26, 2010 (http://www.livablefutureblog.com/2009/03/
    dangerous-medicine-in-us-food-production-antibiotics).

Moss, Michael. "*E. coli* Outbreak Traced to Company That Halted
    Testing of Ground Beef Trimmings." *New York Times*,
    November 12, 2009. Retrieved February 26, 2010 (http://www.
    nytimes.com/2009/11/13/us/13ecoli.html?fta=y).

Moss, Michael. "*E. coli* Path Shows Flaws in Beef Inspection." *New
    York Times*, October 3, 2009. Retrieved February 26, 2010 (http://
    www.nytimes.com/2009/10/04/health/04meat.html?_r=1).

Moss, Michael. "Safety of Beef Processing Method Is Questioned."
    New York Times, December 30, 2009. Retrieved February 26,
    2010 (http://www.nytimes.com/2009/12/31/us/31meat.
    html?scp=2&sq=&st=nyt).

Moss, Michael. "Senate Bill Would Require *E. coli* Testing." *New York
    Times*, November 18, 2009 (http://www.nytimes.com/2009/11/19/
    health/19beef.html?fta=y).

Neuman, William. "After Delays, Vaccine to Counter Bad Beef Is
    Being Tested." *New York Times*, December 3, 2009. Retrieved
    February 26, 2010 (http://www.nytimes.com/2009/12/04/
    business/04vaccine.html?_r=2&fta=y).

Rangel, Josefa M., Phyllis H. Sparling, Collen Crowe, Patricia
    M. Griffin, and David L. Swerdlow. "Epidemiology of
    Escherichia coli O157:H7 Outbreaks, United States 1982–2002."

Centers for Disease Control and Prevention, July 6, 2005. Retrieved February 26, 2010 (http://www.cdc.gov/ncidod/eId/vol11no04/04-0739.htm).

Saldmann, Frederic. *Wash Your Hands! The Dirty Truth About Germs, Viruses, and Epidemics—and the Simple Ways to Protect Yourself in a Dangerous World*. New York, NY: Weinstein Books, 2007.

Salmon, Terry. "Food Safety and Salinas Valley Crops: Rodent Control in Leafy Greens Production." *Western Farm Press*, June 7, 2008. Retrieved February 26, 2010 (http://westernfarmpress.com/mag/farming_food_safety_salinas).

Science Daily. "How *E. coli* Becomes Resistant to Many Antibiotics." April 12, 2007. Retrieved February 26, 2010 (http://www.sciencedaily.com/releases/2007/04/070411073539.htm).

Welse, Elizabeth. "New Methods Aim to Keep *E. coli* in Beef Lower All Year." *USA Today*, February 8, 2010. Retrieved February 26, 2010 (http://www.usatoday.com/news/health/2010-02-08-beeftech08_CV_N.htm).

World Health Organization. "Enterohaemorrhagic *Escherichia coli* (EHEC)." May 2005. Retrieved February 26, 2010 (http://www.who.int/mediacentre/factsheets/fs125/en).

INDEX

ABOUT THE AUTHOR

Kristi Lew is the author of more than thirty science books for teachers and young people. A former high school science teacher with degrees in biochemistry and genetics, she now makes a living writing articles and books designed to educate children and adults about science, health, and the environment.

PHOTO CREDITS

Cover, pp. 8, 26 Justin Sullivan/Getty Images; pp. 4–5, 18, 30–31, 33, 35, 42–43, 45 © AP Images; p. 7 CDC/Peggy S. Hayes, photo by Elizabeth H. White, M.S.; p. 12 Bloomberg via Getty Images; p. 15 Dr. Gary Gaugler/Photo Researchers, Inc.; p. 20 Shutterstock.com; p. 23 Alex Wong/Getty Images; p. 38 Ryan Kelly/Congressional Quarterly/ Getty Images; p. 47 Eric O'Connell/Taxi/Getty Images; pp. 48–49 Andy Sotiriou/Photodisc/Getty Images; interior graphics © www. istockphoto.com/Chad Anderson (globe), © www.istockphoto.com/ ymgerman (map), © www.istockphoto.com/Brett Lamb (satellite dish).

Photo Researcher: Peter Tomlinson